# YOUR KNOWLEDGE HAS VALUE

- We will publish your bachelor's and master's thesis, essays and papers

- Your own eBook and book - sold worldwide in all relevant shops

- Earn money with each sale

Upload your text at www.GRIN.com
and publish for free

# Fabrication of a Solar Cell from Thin Films using Solution Techinique

Cliff Orori Mosiori
Walter Kamande Njoroge
John Okumu

**Bibliographic information published by the German National Library:**

The German National Library lists this publication in the National Bibliography; detailed bibliographic data are available on the Internet at http://dnb.dnb.de.

ISBN: 9783656691587
This book is also available as an ebook.

Print and binding: Books on Demand GmbH, Norderstedt, Germany
Printed on acid-free paper from responsible sources.

The present work has been carefully prepared. Nevertheless, authors and publishers do not incur liability for the correctness of information, notes, links and advice as well as any printing errors.

GRIN web shop: https://www.grin.com/document/275807

# Cd$_x$Zn$_{1-x}$ S/Pbs Thin Film Solar Cell

Mosiori Cliff Orori, Njoroge Walter Kamande, Okumu John

Department of Physics, Kenyatta University, P. Box 43844 -00100, Nairobi, Kenya

## ABSTRACT

In this work, $n$-type Cd$_x$Zn$_{1-x}$S and $p$-type PbS layers were optimized through chemical solution technique for solar cells. Cd$_x$Zn$_{1-x}$S was grown at 82$^0$ C while PbS was grown at room temperature utilizing aqueous conditions. Optical constants suitable for solar cells fabrication were investigated. Cd$_x$Zn$_{1-x}$S films had a band gap varying from 2.47eV ($x = 0.6$) to 2.72 eV ($x = 1.0$), transmittance above 79% in VIS - NIR region with resistivity range of $9.5 \times 10^1$ to $1.22 \times 10^2$ $\Omega$-cm. PbS had a band gap of 0.89 eV, transmittance below 55% with a resistivity range of $6.78 \times 10^3$ to $1.26 \times 10^4$ $\Omega$-cm appropriate for solar cell absorber layers. Their solar cell had a short circuit current, I$_{sc}$ = 0.031A, open voltage, V$_{oc}$ = 0.37 V, efficiency, $\eta$ = 0.9% with a fill factor, ff = 0.66.

## 1. INTRODUCTION

Thin films can be used to fabricate photovoltaic devices that can convert solar energy into electrical energy for various uses. This can be achieved if their structure, inter-band transitions and other optical properties are maximized to harvest enough solar radiation to provide energy. Activities that take place when electrons transit between energy bands in thin films are of fundamental importance in harvesting solar radiation (Chapin and Pearson, 1954). Solar energy is abundant but it has not been harvested well although harvesting it by use of solar cells does not require sophisticated and expensive facilities. Thin film nanotechnology has fabricated cheap photovoltaic cells that produce power for homes, small commercial uses or electric current (Schroder, 1998). Solar energy if a form of energy that is reliable, easy to maintain, install and it can extend to cover IR and UV radiation regions to make solar cells more efficient (Armin, 2009; Siu and Kwok, 1978). Solar cells absorb photons from the sun and convert them directly into electricity. They consist of a $p$-type layer which has a majority hole carriers and an $n$-type layer that has a majority electron carriers. When a photon with energy greater than the band-gap of the semiconductor passes through such a cell, it may be absorbed by the material and this takes the form of a band-to-band electronic transition producing an electron-hole pair.

Several thin films have been fabricated and used to manufacture solar cells using various methods of preparation for solar cells that include chemical deposition; liquid deposition and chemical vapour among others have also been used. Many researchers have devoted their efforts to solution technique because it is a non-expensive method for thin film preparation. It has developed many noble materials. Doping using elemental dopants like boron (Khallaf *et al.*, 2009), indium (Shadia *et al.*, 2008), arsenide and chlorine (Amanullah *et al.*, 2005) has produced suitable films for use as window layers. Ternary derivatives of CdS have generated a lot of research interest because of their varied applications in the field of optoelectronic devices. $Cd_{1-x}Zn_xS$ which is gaining prominence as a good candidate for wide band gap materials for solar cells. Its band gap can be tailored to vary from 2.43 eV to 3.32 eV depending on its constituents and preparation techniques. The addition of Zn onto CdS enhance open-circuit voltage ($V_{oc}$) and short-circuits current ($I_{sc}$) in hetero-junction devices and result into a decrease in window absorption losses. Lead chalcogenides films grown by CBD possess a well-defined band structure in which their energy gap varies continuously between 0.41 eV to 2.7 eV depending on the method of preparation. Since their band gap ranges within the optimum theoretical band gap for maximum absorber material of about 1.5 eV, they can be used to fabricate solar cells (Popa *et al.*, 2006).

## 2. Experimental details

### 2.1 Preparation of substrates

Glass slides were used as substrates, degreased in hydrochloric acid for 24 hours, washed with detergent, rinsed in distilled water and dried in air for 2 hours. They were inserted suspended vertically from synthetic foam which covered the beakers containing the bath solution.

### 2.2 Growth of Cadmium Zinc Sulphide layer

The bath composed of 0.038 M cadmium nitrate, 0.076 M ammonium nitrate, and 0.076 M thiuorea in de-ionized water. 25 ml of the each solution was taken into a separate beaker and de-ionized water added to top up to 100 mL, then heated to about 82° C. Using a burette, $NH_4OH$ (29.4%) was added drop-wise to maintain a pH of about 9. Glass substrates were inserted for 25 minutes. Varying concentrations of zinc nitrate $[Zn(NO_3)_2]$ solutions were added to vary zinc ions and the value of $x$, (i.e. $x = Zn^{2+}/[Cd^{2+} + Zn^{2+}]$ ) was varied from 1.0 - 0.6 according to the equation $Cd_xZn_{1-x}S$. A reaction mechanism for the formation of CdS was suggested to be as shown in equation 2.1 where cadmium salts in the presence of ammonium hydroxide solution form the complex compounds.

$$NH_3 + H_2O \longrightarrow NH_4^+ + OH^- \qquad (2.1)$$
$$Cd(NO_4)_2 + 4(NH_3) \longrightarrow Cd(NH_3)_4[NO_3]_2$$

$$Zn(NO_4)_2 + 4(NH_3) \longrightarrow (Zn(NH_3)_4)[NO_3]_2$$

$$Cd(NH_3)_4^{2+} \longrightarrow Cd^{2+} + 4NH_3 \qquad (2.2)$$

$$Zn(NH_3)_4^{2} \longrightarrow Zn^{2+} + 4NH_3 \qquad (2.3)$$

$$Cd(OH)_2 \longrightarrow Cd^{2+} + 2(OH)^- \qquad $$

$$Zn(OH)_2 \longrightarrow Zn^{2+} + 2(OH)^- \qquad (2.4)$$

Thiourea is a sulphide-ion source in an alkaline medium where the sulphide ions are released slowly as shown in equation 2.5 and 2.6. Cadmium ions react with sulphide and zinc ions to form $Cd_xZn_{1-x}S$ as shown in equations (2.3), (2.4), and (2.6):

$$(NH_2)_2CS + 3OH^- \longrightarrow NH_3 + CO_3^{-2} + S \qquad (2.5)$$
$$SH^- + OH^- \longrightarrow S^{-2} + H_2O \qquad (2.6)$$

$$x\,Cd^{+2} + (1-x)\,Zn^{2+} + S^{-2} \longrightarrow Cd_xZn_{1-x}S \qquad (2.7)$$

The reactions in equations 2.1 to 2.7 are interrelated. Ammonia concentration affects the concentration of cadmium ions $Cd^{+2}$, precipitation of cadmium hydroxide $[Cd(OH)_2]$, concentration of the tetra-ammine-cadmium complex ions $[Cd(NH_3)_2^{+2}]$, and the concentration of hydroxide ions $[OH^-]$ in the bath. Electrical properties were studied using a four point probe connected to Kethley 2400 source meter interfaced with a computer using Labview program while optical transmittance was measured using UV-VIS-NIR spectrophotometer 3700.

3

## 2.3 Growth of PbS thin film layer

Growth was done in a reactive chemical bath in a 100 ml beaker by sequential additions of solution of 5 ml of 0.5 M lead nitrate as a source of $Pb^{+2}$, 5 ml of 2 M sodium hydroxide as source of alkaline medium, 6 ml of 1 M thiourea as source of $S^{-2}$ and 2 ml of 1 M tri-ethanolamine as a complexing agent. Solutions were prepared from analytical grade chemicals. Lead ions of varying concentration from 0.3 M to 0.7 M at intervals of 0.1M were prepared. 5ml of lead nitrate poured into a 100ml beaker followed by 5ml of 2M sodium hydroxide and the mixture was thoroughly stirred using an electric stirrer to obtain a milky solution. This was followed by adding 6ml of 1M thiourea followed immediately by 2ml of 1M tri-ethanolamine while stirring continued for about two minutes to ensure uniformity of the mixture. A substrate was inserted vertically leaning on the side of the beaker and maintained at room temperature for 120 minutes. Lead nitrate concentration was varied at intervals of 0.1 for the subsequent films from 0.3 - 0.7M of $Pb^{+2}$ ion concentrations. Reflection and transmission spectra were measured at room temperature in the spectral range of 260 – 2000 nm (4.54 –1.08 eV) using NIR-VIS IR spectrophotometer DUC 3700 instrument at ambient temperature. The electrical resistivity measurements were done using the Keithley 2400 source meter interfaced with a computer

## 3 RESULTS

### 3.1 Electrical Properties

The resultant PbS thin films were homogeneous, well adhered to the substrate and specularly reflecting. As shown in figure 1, resistivity of PbS was $9.171 \times 10^3$ $\Omega$-cm and decreased to $6.78 \times 10^3$ $\Omega$-cm but thereafter increased almost linear to $1.26 \times 10^4$ $\Omega$-cm at 0.7M PbS which translates to an electrical conductivity range of $1.09 - 0.79 \times 10^{-5}$ S-cm$^{-1}$. This observation was attributed to the large levels of scattering centres due to amorphous nature of the films as lead concentration increases. Amorphous thin films have higher concentration of scattering centres.

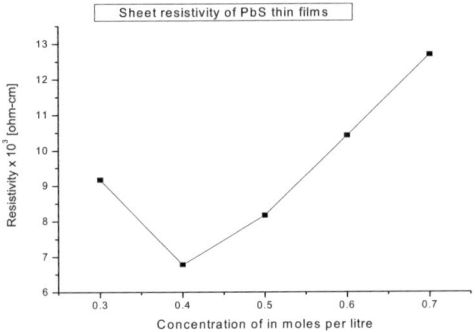

**Figure 1**: *Graph of resistivity against concentration of PbS films*

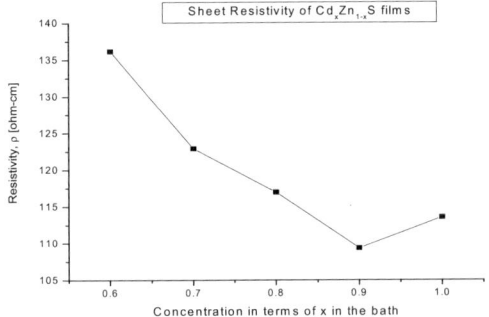

**Figure 2**: *Optical properties of $Cd_xZn_{1-x}S$*

On the hand, electrical resistivity of $Cd_xZn_{1-x}S$ increased with increase of Zn ions from $1.09 \times 10^2$ to $1.36 \times 10^2$ Ω-cm shown in figure 2. Introduced Zn impurities result into more scattering centres which in turn reduced the mean free path hence scattering at the grain boundaries in addition to the bulk scattering centres that were present. Charge carries moving through the thin film grains suffer extra scattering at the grain boundaries hence a reduced mean free path. Tuning the electrical properties of the thin films through doping is very important for various applications. Similar sheet resistivity measurement have been observed (Vidhya and Velumani, 2009, Song *et al.,* 2006) by Van der Pauw technique {order of $10^1$ Ω-cm-$10^2$ Ω-cm} or {conductivity order of $1 \times 10^{-2}$ to $7 \times 10^{-3}$ $[\Omega\text{-cm}]^{-1}$} as compared to conductivity obtained that ranged from $7.4 \times 10^{-3}$ to $9.1 \times 10^{-3}$ S/cm.

## 3.2 Optical Properties

$Cd_xZn_{1-x}S$ films were smooth, uniform, adherent, bright yellow orange in colour where the yellowness decreased with increasing zinc content. Zn reduced reflectance and absorbance while it increased transmittance within the visible and infrared range. Refractive index (n) was calculated using the equation proposed by Ravindra *et al.* (2006) as;

$$n = [(1+R^{1/2})/ (1-R^{1/2})] \qquad (3.4)$$

while optical refractive index (n) and energy band gap ($E_g$) related as;

$$n = 4.08 - 0.62E_g \qquad (3.5)$$

***Table 1:*** *Optical and electrical properties of $Cd_xZn_{1-x}S$*

| Film | Conc. of *Zn* | [n] | $E_g$ [eV] | $\rho$ [$\Omega$-m] |
|------|------|------|------|------|
| $Cd_{06}Zn_{04}S$ | 0.4 | 2.39 | 2.72 | 136.19 |
| $Cd_{07}Zn_{03}S$ | 0.3 | 2.41 | 2.69 | 122.91 |
| $Cd_{08}Zn_{02}S$ | 0.2 | 2.46 | 2.60 | 116.98 |
| $Cd_{09}Zn_{01}S$ | 0.1 | 2.51 | 2.52 | 109.37 |
| CdS | 0.0 | 2.53 | 2.47 | 113.56 |

Refractive index decreased with increase in Zn concentration and this explains why the colour of the films faded as concentration increased. Dielectric constants are used to describe any losses caused by optical conductivity ($\sigma$) in thin films where real and imaginary parts of the dielectric constant are given by;

$$\varepsilon_c = \varepsilon_r + \varepsilon_i \qquad (3.6)$$

and they were estimated using the relations;

$$\varepsilon_1 = n^2 - k^2 \qquad (3.7)$$

$$\varepsilon_2 = 2nk \qquad (3.8)$$

6

The dielectric constant ($\varepsilon$) reduced as wavelength increased at a constant Zn concentration and it explains why there are high optical conductivity loses at longer wavelengths. The absorption coefficient ($\alpha$) was calculated using the equation;

$$\alpha = 2.303A/d \qquad (3.9)$$

where $\alpha$ is the absorbance coefficient value at a particular wavelength($\lambda$) and $d$ is the thickness of the semiconductor film. Extinction coefficient on the other hand was calculated using the relation;

$$k = \alpha\lambda/4\pi \qquad (3.10)$$

Both coefficients were very small over a wide range of wavelengths and as such very low photon energy absorption losses are experienced. They form high quality window layer materials (Kumar and Sankaranayanan, (2009), Kasim *et al.* (2008), Saliha, (2009), Vidhya and Velumeni, (2009) with transmittances of 90%, 80%, 79% and 65% respectively in the wavelength range of 300 – 1200 nm.

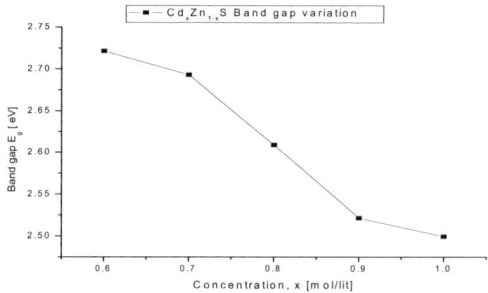

***Figure 3****: Band gap variation for $Cd_xZn_{1-x}S$*

### 3.3 $Cd_xZn_{1-x}S$ / PbS Solar cell

A layer of $Cd_xZn_{1-x}S$ was chosen as the window layer with the largest band gap of 2.72 eV, highest transmittance of above 79%, lowest refractive index, lowest optical absorption, extinction coefficient. PbS with lowest refractive index, lowest transmittance of below 55%, and high absorption and extinction coefficients as an absorber layer. Silver paste was used as the ohmic contact. Solar cell's I-V characteristics were measured using a solar simulator (table 2). Figure 3 illustrates current/or and power variation with voltage. Table 2 display experimental I-V characteristics of the solar cell while other cell parameters are presented in table 4.

**Table 2:** *Variation of current against voltage in $Cd_xZn_{1-x}S/PbS$ Solar cell as measured by solar simulator.*

| Voltage [V] | Current [A] | Power [w] |
|---|---|---|
| 0.00 | 0.031 | 0.0 |
| 0.05 | 0.031 | 0.00155 |
| 0.10 | 0.031 | 0.0031 |
| 0.15 | 0.031 | 0.00465 |
| 0.20 | 0.030 | 0.006 |
| 0.25 | 0.028 | 0.007 |
| 0.27 | 0.027 | 0.00729 |
| 0.29 | 0.026 | 0.00754 |
| 0.31 | 0.023 | 0.00713 |
| 0.33 | 0.020 | 0.0066 |
| 0.35 | 0.015 | 0.00525 |
| 0.36 | 0.010 | 0.0036 |
| 0.37 | 0.001 | 0.00037 |

**Table 3:** *Cell parameters of $Cd_xZn_{1-x}S/PbS$ Solar cell*

| Cell parameters | Value of parameter/unit |
|---|---|
| $I_{sc}$ | 0.031   A |
| $V_{oc}$ | 0.37   V |
| $V_{max}$ | 0.29   V |
| $I_{max}$ | 0.026   A |
| $P_{max}$ | 0.00754 W |
| $ff$ | 0.66 |
| $\eta$ | 0.9 |

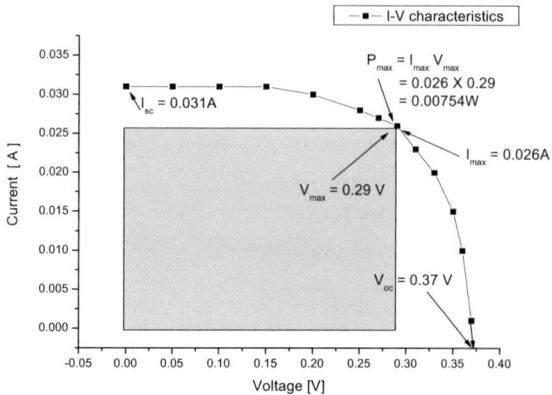

*Figure 4: I-V curve characteristics of $Cd_xZn_{1-x}S$ /PbS solar cell*

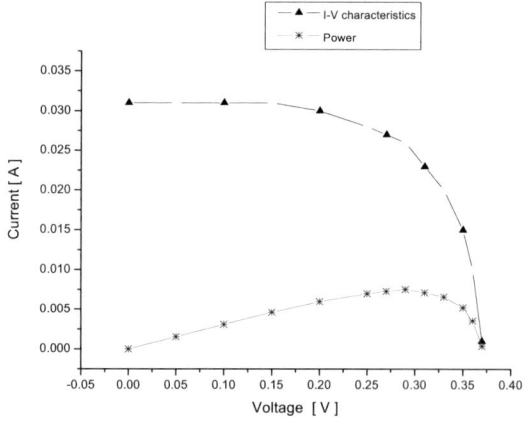

*Figure 5: I-V curve and power-voltage characteristics of $Cd_xZn_{1-x}S$ /PbS Solar cell*

The solar cell had a short circuit current, $I_{sc}$ = 0.031 A, open voltage, $V_{oc}$ = 0.37V, a fill factor, ff = 0.66 or 66 % and an efficiency, $\eta$ = 0.9 as shown in Table 4. The cell produced a higher current than what was reported by Harumi *et al.* (1995) with a $V_{oc}$ of 0.45V and $I_{sc}$ of $0.001mA/cm^2$. Fill factor is a measure of the quality of a cell and therefore a large fill factor of 1 is desirable and it corresponds to I-V sweep that is more square-like where typical fill factors ranges from 0.50 to 0.82 (50 % - 82%). Considering the square-like nature of the I-V curve and a fill factor of 0.66, it was concluded that the cell formed a good $p$-$n$ junction for solar cells.

## 4. CONCLUSIONS

$Cd_xZn_{1-x}S$ and PbS thin films were successfully prepared on glass substrates using CBD technique under different preparation conditions. Optimum conditions gave $Cd_xZn_{1-x}S$ and PbS solar cell that gave a short circuit current, $I_{sc}$ = 0.031 A, open voltage, $V_{oc}$ = 0.37 V, a fill factor, ff = 0.66 or 66 % and an efficiency, $\eta$ = 0.9. The solar cell had a fill factor above 0.5 (i.e. ff = 0.66) support a good p-n junction for solar applications.

## ACKNOWLEDGMENTS

We acknowledge the Department of Physics, Kenyatta University where this study was carried out and the Department of Material Science University of Nairobi, [Chiromo Campus] for kindly providing the UV-VIS-NIR spectrophotometer 3700 and the Solar Simulator for analysis.

## REFERENCES

Amusan, J., Fajinmi, G. and Sasusi, Y. (2007). *Deposition time dependence on absorptivity of chemically deposited lead sulphide thin films*. Research Journal of Applied Sciences, **9**: 931-937.

Bacaksi, E. and Cevik, U. (2006). *K-shell fluorescence yield of Cd and Zn in $Cd_{1-x}Zn_xS$ thin films on the structural, optical and electrical properties of MW-CBD CdZnS thin films*. American Journal of Nanotechnology, **123**: 1-5

Chouldrury, N. and Sarma, K. (2008). *Structural characterisation of nanocrystralline PbS thin films: synthesised by CBD method*. Indian J. of Pure and Appl. Phys, **46**: 261-265.

Dzhafarov, T., Ougul, F. and Karabay, I. (2006). *Formation of CdZnS thin films by Zn diffusion*. J. Phy. D: Appl. Phy. **39**: 3221-3225

Eshafie I. and Ekpunobi, A. (2004). *Optical properties and band offsets of CdS/PbS superlattice*. The Pacific Journal of Science and Technology, **11**: 404-407.

Gaewdang, N., Gaewdang, T. and Lipar, W. (2004). *Some characterisation of chemical bath co-deposited CdS/ZnS thin films*. Technical Digest Inter. PVSEC, **14**: 124 -126

Ghamsari, M. and Khosravi, A. (2005). *The influence of hydrazine hydrate in the preparation of lead sulphide thin film.* Iranian J. of Sci. & Technology, **29:** 151- 162. Journal of Science and Technology, **8:** 155-161

Kasim, U., Narayanan H. and Anthony, O. (2008). *Optimization of process parameters of chemical bath deposition of Cd₁₋ₓZnₓS thin films.* Leonardo Journal of Sciences, **12:** 111-120.

Kumar, T. and Sankaranarayanan, S. (2009). *Growth and characterisation of CdZnS thin films by short duration microwave assisted chemical bath deposition technique.* Chalcogenide Letters, **6:** 555- 562

Mulik, R., Pawar, S., More, P., Pawar, A. and Patil, V. (2010). *Nanocrystalline PbS thin films: synthesis, microstructural and optoelectronic properties.* Scholars Research Library, **2:** 1-6.

Osherov, A., makai, J., Balasz, J., Horvath, Z., Gutman, N., Amir, S. and Golan, Y. (2010). *Tenability of optical band edge in thin PbS films chemically deposited on GaAs (100).* Journal of Condensed Matter, **22:** 1-7.

Oztas, M. and Bedir, M. (2005). *Some Properties of Cd₁₋ₓZnₓS and Cd₁₋ₓZnₓS(In) Thin Films Prepared by Pyrolytic Spray Technique.* Journal of Applied Sciences, **4:** 534-537.
Patil, R., Lisca, M., Stancu, V., Buda, M., Pentia, E. and Botila, A. (2006). *Crystalline size effect in PbS thin films grown on glass substrates by chemical bath deposition.* Journal of Optoelectronics and Advanced materials, **1:** 43-45.

Pentia, E., Pintilie, L., Matei, I., Botila, T. and Ozbay, E. (2001). *Chemical prepared nanocrysalline PbS thin films.* Journal of Optoelectronics and Advanced Materials, **3:**525 – 530.
Popa, B., Ray, S. and Barma, A. (2006). *Properties of chemically deposited PbS thin films.* Jorn. of Appl. Physi. **21:** 43-45.

Popescu, V., Jumate, N., Popescu, G., Moldovan, M. and Prejmeriean, C. (2010). *Studies of some electrical and photoelectrical properties of PbS films obtained by sonochemical methods.* Chalcogenide Letters, **7:** 95-100.

Ravindran, N., Ganapathy, P. and Choi, J. (2006). *Energy gap refractive index relations in semiconductors – An overview.* Elsevier Thin Solid Films: Infrared Physics and Technology, **50:** 21-79

Saliha, I., Muhsin, Z., Yasemin, C. and Mujdat, C. (2009). *Optical characterisation of the CdZn(S₁₋ₓSeₓ)₂ thin films deposited by spray pyrolysis method.* Optica Applicata, **XXXVI:**1-9.

Seghaier, S., Kamoun, N., Brini, R. and Amara, A. (2006). *Structural and optical properties of PbS thin films deposited by chemical bath deposition.* Elsevier Materials, Chemistry and Physics, **97:** 71-80.

Song. J., R. Thapa, R., Maity, K. and Chattopadhyay K (2005*) Optical and dielectric properties of PVA capped nanocrystalline PbS thin films synthesized by chemical bath deposition* Thin Solid Films, **211:** 14–21

Ubale, A., Junghare, A., Wadibhasme, N., Darypurkar, A., Mankar, R. and Sangawar, V. (2007). *Thickness Dependent Structural, Electrical and Optical Properties of Chemically Deposited Nanopartical PbS Thin Films.* Turk. J. Phys., **31**: 279 – 286.

Valenzuela, J., J´auregui, R., Ram´ırez-Bon, A., Mendoza-Galva´n, M. and Sotelo-Lerma, A. (2003). *Optical properties of PbS thin films chemically deposited at different temperatures.* Thin Solid Films, **441**: 104–110.

Vidhya, Y., and Velumin, A. (2009). *Electrical characterisation of chemically deposited thin films under magnetic field.* Phys. Stat. Sol. **167**: 143-151

**Preliminary Bibliography**
Mr. Mosiori Cliff Orori is lecturer at Rift Valley Institute of Science and Technology-Nakuru. His research interests are in Material Science for electronic devices. He is associated with the Solid State Physics group of Kenyatta University. Has a wide knowledge Electronics lab, Digital Logic Lab, Microprocessor systems, Electrical Circuits, Data communications, Instrumentation and Control systems. He is a PhD student at Kenyatta University currently working on his PhD proposal entitled *"Effect of Silver Nanoparticles on optical properties of cerium titanium oxide thin films for UV-protection coatings on optical fibres"* deposited by chemical bath deposition. **Dr. Njoroge Walter Kamande** is a senior lecturer and Chairman of Physics department of Kenyatta University, Kenya. **Prof. Okumu John** is a professor of Physics and serves as the DVC academic at Kenyatta University, Kenya.

# YOUR KNOWLEDGE HAS VALUE

- We will publish your bachelor's and master's thesis, essays and papers

- Your own eBook and book - sold worldwide in all relevant shops

- Earn money with each sale

Upload your text at www.GRIN.com
and publish for free